Power of Scotland
The Nation's Generation Stations
Guthrie Hutton

Men, some wearing caps fitted with miners' carbide lamp holders, pictured at one of the shaft entrances to the tunnel they were driving through Ben Nevis to the power station at the British Aluminium Company's Lochaber smelter.

The power station of the Loch Sloy Hydro Electric Scheme situated at Inveruglas on the shores of Loch Lomond.

Acknowledgements

With some power stations consigned to history, the aim of this book is to recognise the electricity supply industry's achievements, before memory fades completely. In compiling the images and stories I must thank Margeorie Mekie, Allan Thomson and Robert Hunter for valuable information and pictures. I must also thank George Nairn for the pictures by Dunfermline photographer Peter Leslie on pages 38 and 39. I am grateful too to Fife Cultural Trust (on behalf of Fife Council) Dunfermline Carnegie Library and Galleries for the pictures on pages 17 and 43. The picture on page 42 is reproduced courtesy of the National Mining Museum Trust. The picture on page 19 is reproduced by permission of East Ayrshire Leisure / East Ayrshire Council.

First published in the United Kingdom, 2019,
by Stenlake Publishing Ltd.
www.stenlake.co.uk
ISBN 978-1-84033-852-2

Printed by
P2D Books, 1 Newlands Rd, Westoning, Bedford, MK45 5LD

Further Reading

The following were the principal books used by author during his research. Only *Falkirk's Trams and Early Buses* is availble from Stenlake Publishing; for the rest please contact your local bookshop, reference library or search for them on the internet.

Brotchie, Alan W., *Falkirk's Trams and Early Buses*, 2011.
Brotchie, Alan W., *The Tramways of Falkirk*, 1975.
Campbell, Patrick, *Tunnel Tigers*, 2005.
Corporation of Glasgow, *Municipal Glasgow: Its Evolution and Enterprises*, 1914.
Dick, David, (Editor), *A Scottish Electrical Enlightenment*, 2000.
Hood, John, *John Brown Engineering: Power Contractors to the World*, 2004.
Jay, Kenneth, *Calder Hall*, 1956.
Miller, James, *The Dam Builders*, 2002.
Royal Philosophical Society of Glasgow, Proceedings *Vol. LVIII*, 1930.
SSEB News pages 6 & 7, Report on lecture by John S. Forrest, Nov. 1975.
Wood, Emma, *The Hydro Boys*, 2002.
Booklets and leaflets:
North of Scotland Hydro-Electric Board, *The Tummel Valley Hydro-Electric Schemes*, c.1950.
North of Scotland Hydro-Electric Board, *Carolina Port 'B'*, 1965.
North of Scotland Hydro-Electric Board, *Power From the Glens*, 1973.
Scottish Hydro Electric (SSE), *Power From the Glens* 2005.
SSEB, *Hunterston A Nuclear Power Station, c.*1969.
SSEB, *Barony Generating Station*, 1957.
SSEB, *Cockenzie Power Station.*
SSEB, *Longannet Power Station.*
Scottish Power, *Methil Power Station, c.*1995.

Introduction

In this gadget-filled modern world, it's easy to take electricity for granted, but plugging in or switching on would be futile unless something was generating power at the other end of the wire. Electricity has always been there of course, with the natural world's most obvious manifestation being a bolt of lightning, but while such bangs and flashes terrified some of our forebears, others tried to find out what caused them. As is often the case in science the significance of some early discoveries was not understood while others occurred almost by accident, as by-products of experiments that sought to prove something else. Knowledge inched forward, the dynamo was developed and then, in the 1870s, when Joseph Swan and Thomas Edison both independently invented the electric filament lamp, people literally saw the light.

The first systems were based on Thomas Edison's preference for low voltage direct current (DC), but another American inventor, Nicolas Tesla, began to champion the use of high voltage alternating current (AC). Regarded as more hazardous, it could nevertheless be transmitted over longer distances and needed fewer, larger power stations. Tesla eventually won the war of ideas and AC was adopted universally.

In Britain the electricity supply industry began with the passing of the Electric Lighting Act in 1882, followed by a rush of applications to the Board of Trade from people keen to take advantage of the new source of energy. Big cities, large burghs, a number of tramway systems and some industrial users were soon authorised to supply and distribute electricity within prescribed boundaries. The number of power stations grew, but they varied in size and used different voltages until, in 1926 government intervened to regulate the industry, standardise supply and develop a transmission grid. Many small operations were decommissioned in favour of fewer, larger stations and in 1948 the industry was nationalised, managed by area boards. That structure was reversed in 1989 when a new parliamentary act paved the way for privatisation, enacted in Scotland in 1991.

Initially coal was the only fuel and it grew in importance as demand increased and generating stations, sited close to the coalfields got bigger. Water-driven systems using the flow of rivers or high-head reservoirs were developed and became almost commonplace in parts of Scotland. Viewed through opposing arguments as either the ideal renewable or environmentally damaging, hydro-electricity remains a valuable source of power, although only one big conventional high-head scheme, the 100MW Glendoe station beside Loch Ness, has been built since the 1950s. Nuclear power, which seemed to hold the promise of long-term cheap power, was introduced in the 1950s, but fell out of favour as people grew wary of the potential hazards. Diesel generators served small communities and, until the costs began to rise, oil and gas seemed like the future. Since then concerns over climate change and global warming have led to a retreat from fossil fuels and development of a range of renewable technologies: solar panels, wind turbines, biomass and systems to harness waves or tides have all been designed and other ideas, to generate or store power in large quantities, may be waiting round the corner to be discovered. The electricity supply industry has thus continuously adapted to meet demand and after lighting the first bulb, has never stopped generating.

Glasgow Corporation's St. Andrew's Cross Power Station

In 1890 Glasgow Corporation obtained an order to provide electricity throughout the city, basing its first supplies on small operations installed by the private electrical engineering firm of Muir, Mavor and Coulson. Thanks to them the General Post Office in George Square was the first in Britain to be lit with electric light and later, as Mavor and Coulson, the firm became better known for the development of mining equipment and electrical installations in mines. The Corporation built power stations, at John Street, Waterloo Street (later converted to a sub-station) Port Dundas and St Andrew's Cross. They took over the operations of other private suppliers in the city and absorbed the electrical supply undertakings of the burghs of Govan and Partick when the city's boundaries were extended in 1912. The pictures show (above) the Govan station, which had been in operation since 1900 and Port Dundas (right).

The Port Dundas generating station was situated on a four and half acre site alongside South Speirs Wharf on the Forth & Clyde Canal. The canal was important: the water was used for condensing purposes and barges delivered coal, unloaded directly into the station by a travelling crane. The station was built in three sections, each designed to accommodate steam turbines and generating plant of 10,000 horsepower. All three sections had a large chimney and in the early days of broadcasting these served a double purpose, with wires stretched between them as a radio transmitter for the infant British Broadcasting Company, the forerunner of the BBC. Following the station's decommissioning and demolition in the late 20th century BBC Scotland acquired the land as a base for outside broadcast vehicles, but it was never built. Later, Scottish Opera set up a technical and production facility on the site, so the one-time generating station has also had a long association with arts and the media.

Just around the corner from Port Dundas was the huge Pinkston Power Station. For the city to have two large generating stations within a few hundred yards of each other might seem excessive, but Pinkston was not intended for general use: it had the single dedicated purpose of generating power for the city's tram system. Many of the tramway systems that were built in the early 20th century to serve towns and cities had their own dedicated generating stations, but Pinkston was by far the biggest, because Glasgow's tram system was the biggest (and some would argue, the best!). Glasgow Corporation took over the tram systems from the earlier private operators in 1894. At that time the trams were horse-hauled, but after a couple of trials the Corporation resolved to go electric and by April 1902, the horse cars had gone.

The Pinkston power station was built on the site of a former pig iron store beside the Forth & Clyde Canal's Port Dundas basins. As at Speirs Wharf, canal water was used for condensing purposes and barges could come alongside with cargoes of coal. The Caledonian and North British Railway companies also served the station. Driven by steam turbines, Pinkston's generators produced three-phase alternating current at a high voltage, which was transformed into continuous 500 volt current at five sub-stations around the city; Coplawhill, Kinning Park, Partick, Dalhousie, and Whitevale, with a sixth at Rutherglen added later. With the contraction and eventual demise of Glasgow's trams, Pinkston became a public supply station. A huge concrete cooling tower, built alongside in the early 1950s, was demolished with explosives in 1976.

Demand for electricity grew rapidly and within a decade of starting up their new power stations, Glasgow Corporation realised that more capacity was needed and just prior to the First World War secured a site for a new station adjacent to Dalmarnock Bridge on the River Clyde. The planned station would be huge by comparison with what had gone before, but delayed by the war it didn't start generating until 1920. By that time the industry was in need of restructuring. Since 1882 when the first Electric Lighting Act was passed, hundreds of public and private suppliers had been authorised, all operating to their own standards. It was a mess. The government set up the Electricity Commission in 1920, but with no statutory powers it made little difference. Five years later the Weir Committee was appointed to consider the options and following their report, the Electricity (Supply) Act of 1926 was passed, starting a process that shaped the modern industry.

The 1926 act established the UK wide Central Electricity Board, which was set up in March 1927. It acted as a wholesaler, buying power from the former suppliers who bought it back as retailers and sold it on to their customers. Power stations were to be interconnected by a grid of transmission lines, an arrangement that demanded uniform standards. Big changes were needed in Glasgow to bring its system into line with other operators, but because government was dictating the change, the city didn't have to pay. Three Scottish distribution areas were created, north, south and central. The number of generating stations in the 'Central Scotland Electricity Scheme' was reduced to ten, with Dalmarnock as the nerve centre. The other selected stations were Clyde's Mill, Yoker, Stonebyres, Bonnington, Portobello, Bonnybridge, Carolina Port (Dundee), Dunfermline and Kilmarnock. The Dalmarnock sub-station is seen here in the late 1920s.

As the centre of population and industry Glasgow had a huge appetite for electricity, but occupied a relatively small geographical area surrounded by other heavily populated and industrialised areas. The challenge of providing power for these was taken up in the early days by a private concern: the Clyde Valley Electrical Power Company. The company was authorised by Act of Parliament in 1901 and set about building two large power stations one of which was located at Leven Street in Motherwell. The contract for its design and construction was awarded to the British Westinghouse Electrical and Manufacturing Company of Manchester, with the boilers supplied by Babcock and Wilcox. The station is seen here shortly after going into operation in January 1906, with two large wooden cooling towers on the right. Despite regular upgrades, it was not included in the central Scotland scheme of the late 1920s and closed shortly after.

The Clyde Valley Electrical Power Company set up their other large station at Yoker on the north bank of the River Clyde. Designed and equipped by the same contractors who built Motherwell, it came on stream in June 1905. At that time, Yoker was not in Glasgow, but following boundary changes of 1912 came within the city and despite being on a rival operator's territory remained a unit of the private company. It started with two 2,000kW (kilowatt) turbo-alternator sets and was quickly upgraded. Two 20MW (megawatt), sets were installed in 1929 and 1931 and for the next 40 years an almost continuous process of upgrading brought it up to an eventual installed capacity of 100MW remaining in operation until 1976. The station's principal distinguishing features were the towers needed to carry the power lines across the Clyde. The one adjacent to the station was 270 feet high, while on the south side of the river the tower, seen here from the Yoker-Renfrew ferry slip, was 283 feet high, giving a clearance of 228 feet over high water. Additional towers of 218 feet high carried the lines across the River Cart on the Yoker to Greenock route.

Following the First World War, the Clyde Valley Electrical Power Company turned its attention to harnessing the power of the river itself above New Lanark. Hydro-electricity wasn't new, since the late 19th century some country houses had their own private systems as, on a much bigger scale did aluminium smelters, but this was the first use for a public supply. There was an outcry: Cora Linn, Bonnington Linn and Stonebyres Linn, collectively known as the Falls of Clyde were famed for scenic beauty and any diminution of that appalled people. A development subsidiary, the Lanarkshire Hydro-Electric Power Company, was set up and the scheme given parliamentary approval in 1924. There were no reservoirs, this was a 'run-of-river' scheme with tilting weirs designed to extract water, while maintaining the natural flow. The full system, with two power stations, Bonnington and Stonebyres (*above*) was commissioned in 1927 and, automated in 1970 it has stood the test of time.

The Clyde Valley Electrical Power Company's initial plans included a third station at Crookston on the south side of the Clyde. It was never built, but the company did eventually establish a station south of the river at Cambuslang. Known as Clydesmill, the name did not refer to the river, but to Mr. Clyde's Mill that formerly occupied the site, and helpfully came with water rights. The station began generating power in November 1916 with a 6,000kW generator and two boilers. Four more boilers and another 6,000kW generator were installed in 1918, and in 1921 and 1926 two 18,750kW sets were added. Retained as a base-load station, the installed capacity was raised to 79.5MW with the addition of a fifth generator of 30MW in 1935. Like the other Glasgow and Clyde Valley stations, Clydesmill generated its power by burning coal, readily accessed from Lanarkshire collieries and stored evenly beside the power station by this giant travelling telpher bridge. The main part of the station closed in 1978 and was subsequently demolished, although gas turbines installed in the 1960s remained in operation longer with the whole site cleared by the year 2000.

While Glasgow and the Clyde Valley were powering up, Edinburgh's entry into the generating business was more tentative. The city was even beaten to installing electric trams by Leith, which rejected the capital's proposal to establish a cable car system and in 1905 went ahead with its own electric system. In 1920, and to the dismay of its inhabitants, the Burgh of Leith became part of Edinburgh, but the city's boundary extension did allow the dust to be blown off a pre-war plan for construction of a major new coal-fired power station at Portobello. Situated on High Street between King's Road and Rosebank Lane (now Westbank Street) the station was formally opened in 1923 by King George V. It was extended in 1938, with the most conspicuous alteration being to replace six chimneys with a single stack, 365 feet high, so tall that when the station closed in 1977 it had to be taken down, brick-by-brick to prevent damage to adjacent houses. The water in the adjacent outdoor swimming pool also went cold after heat from the power station was turned off!

3750 K.V.A. 3 phase 60 cycle 66,000/22,000 volt
Outdoor Type Transformers on test at our Works.

BRUCE PEEBLES & CO., LTD.
ENGINEERS.
EDINBURGH.

While the civic authorities may have been tardy, one Edinburgh business, that of David Bruce Peebles, was anything but. Starting as a manufacturer of gas meters in the Fountainbridge area in 1866, David Peebles moved ten years later to Leith. A Dundonian, he called his new factory the Tay Works. By the 1890s the company was making electrical equipment and set up a department for this in 1898. David Peebles died the following year, but with his sons, both trained engineers, involved, the business kept going until it was formed into Bruce Peebles & Company Ltd in 1902. While still retaining a gas department the focus shifted to manufacture and supply of electrical installations. A new factory at East Pilton was opened in 1904 and the business grew to become one of the largest manufacturers of electricity generating equipment and transformers in Europe and one of the most respected names in the electrical installations industry. The works were so big that it had an internal railway system – the first electric railway in the city – with locomotives designed and built in the company workshops.

Bruce Peebles & Company was instrumental in setting up the Scottish Central Electric Power Company and building and equipping its Bonnybridge power station, handily placed to provide power to Falkirk. A tramway system for the town had been authorised by parliament in 1901, but failed to attract financial backing until Bruce Peebles stepped in and acquired the rights in May 1904. By the end of the year they had promoted another company to build the system, but hit a snag because Falkirk Burgh already had their own power station, intending that it should be used for the trams. Once the lawyers had resolved the dispute, Bruce Peebles got on with the job; in the space of five years they had not only built a large power station, but also the infrastructure of one of its major customers. That was good business! The last tram ran in 1936, but by that time the Bonnybridge power station's immediate future had been secured as one of the ten in the Central Scotland Scheme. It is seen here shortly after closure in the 1970s.

With the exception of Bonnington and Stonebyres, the country's early power stations were all coal-fired and most large units were built in close proximity to a major coalfield. Dunfermline's power station at Townhill was no exception, almost surrounded by collieries. The boilers in this picture, typical of their kind, are thought to be those at Townhill. The three stokers are also typical of men who spent their days shovelling coal and raking ash: a hot, dirty, backbreaking job. Bruce Peebles were again involved in building the station, this time for the Fife Electric Power Company. It opened in 1905, although construction of the extensive tramway system, which was to be a principal customer didn't get under way until 1909. The trams served the area's mining towns and villages: Cowdenbeath, Crossgates, Kelty, Lochore and Lochgelly as well as the new Rosyth Dockyard, but by 1937 had succumbed to competition from buses. Meanwhile the Townhill plant had been upgraded to one of the ten main Central Scotland generating stations. Redesigned in the mid 1960s as a twin gas turbine station by the South of Scotland Electricity Board (SSEB), it was decommissioned about 20 years later.

Bo'ness on the south side of the Forth estuary was another town with a coal industry on the doorstep handily placed to supply the burgh's electricity works in Main Street. The station opened in February 1905 and civic pride in the event was clearly evident when these councillors and employees posed in front of the generating equipment. The National Electric Construction Company was awarded the contract to build the works, which consisted of a boiler house, with two boilers, an engine room with two generators, test and battery rooms. Bo'ness was typical of many small municipal stations erected throughout the country with street lighting, domestic customers and small businesses as the principal users. The large local collieries, in common with others in the coal industry, generated their own power. As the national emphasis switched to fewer, large interconnected stations, local operations such as Bo'ness were shut down and their customers transferred to the grid.

Kilmarnock, one of the ten power stations chosen by the Electricity Commissioners for inclusion in the Central Scotland Scheme, occupied a key position on the grid connecting central Scotland to Carlisle by way of Dumfries. It was situated to the south of the town on a former mill site beside Riccarton Bridge, which spanned the River Irvine; a valuable source of condensing water. The station opened in 1904, mainly to supply power for the local tram service. Ten years later, Kilmarnock Corporation was authorised to extend the area of supply to a large part of north Ayrshire and over the next few years much of the county came to be served by the station. To meet the demand, an extension, virtually a new station that more than doubled capacity, was opened in 1922, and when Ayr Power Station closed in 1924 Kilmarnock became the county's principal source of power. Starting with an output of 475kW the station's installed capacity was progressively increased to 85MW, but time eventually caught up with it. Two concrete cooling towers, one erected in 1939 the other in 1942, were demolished in 1976 and the site cleared by 1979.

A lot of cheap reliable electricity was needed for aluminium smelting, and the best way to generate it was a dedicated hydro-electric system. The process of isolating alumina (aluminium oxide) by electrolysis was discovered in 1886 and when allied with the abundant mineral, bauxite, in 1892, the stage was set for a new industry. Formed in 1894, the British Aluminium Company (BAC) quickly acquired the patent rights and set about erecting a smelter beside Loch Ness with a power station driven by water from the scenically famous Falls of Foyers. The potential loss of this natural wonder caused an outcry, but the prospect of industrial employment in the Highlands was too good a prize. Demand soon outstripped supply and the BAC decided to build a new smelter at the head of Loch Leven. High in the hills, the Blackwater Dam was built to impound a reservoir from which water was channelled and piped to the power station at the new village of Kinlochleven. Opened in stages, the system was fully operational by 1909. The power station is seen here behind the tailrace with the water, having driven the turbines, roaring back into the River Leven.

Kinlochleven was big, but not big enough and in 1924 work began on a scheme to establish a new smelter at Fort William. It involved using Loch Treig as a reservoir, and diverting the water through a fifteen-mile-long tunnel bored through Ben Nevis (*see page 1*). Eleven shafts were driven into the hillside so that, including the outlet above Fort William the tunnel could be advanced on 23 faces simultaneously. The final barrier between Loch Treig and the tunnel was blasted out in 1929, but that wasn't the end of the scheme. A huge dam was built to raise the level of Loch Laggan and send the water along a three-mile tunnel to Loch Treig where another dam lifted the level by 35 feet. Further work, to divert water from the River Spey floodplain into Loch Laggan, completed the scheme in 1943. The turbine hall at Fort William, seen in this picture, was equipped with twelve machines generating a combined 86MW of power. Compared to this, the system at Foyers that started it all was tiny and it closed in 1967, but remained a valuable asset and was re-engineered as a reversible pump storage hydro-electric station, opening in 1974.

The Galloway Hydro-Electric Scheme in the south west of the country was authorised in 1929 to harness water in the forty-mile long valley formed by the Water of Deugh and the rivers Ken and Dee. A dam was built to contain Loch Doon in southern Ayrshire as the topmost reservoir, with a pipeline from it feeding a channel known as the Carsphairn Lane. The Deugh and Kendoon Dams impounded the next reservoir above the 21MW Kendoon Power Station. Utilising the same infrastructure, the small Drumjohn station, originally a contractor's sub-station, was added in 1985. Next came the 12MW Carsfad Power Station with a 70 feet high, 550 yard long curved mass concrete dam, seen here under construction. The same arch and mass concrete structure was used for all four dams below Loch Doon down to Earlstoun where the next dam and 12MW power station were built near the village of St. John's Town of Dalry. Kendoon, Carsfad and Earlstoun stations were all commissioned in 1936.

The Galloway scheme was mainly run-of-river, but it included one high-head development that operated independently while adding water to the River Ken below Earlstoun. In the hills to the west, a mass concrete gravity dam, 490 yards long and 78 feet high was built across the Blackwater of Dee to create a man-made reservoir. Slightly curved, for aesthetic rather than structural reasons, it also has a wonderfully evocative name: Clatteringshaws Dam. From the reservoir a three and half mile long tunnel, eleven feet in diameter takes the water to the pipeline, which from a head of 365 feet delivers it to the 24MW Glenlee Power Station. Larger than the upstream stations, Glenlee was also commissioned a year earlier, in 1935 and later became the control station for the whole scheme. Having powered the turbines, the water discharges along a lengthy tailrace to the River Ken, adding to the generating capacity of the final station in the scheme.

Further south, a large barrage was built across the River Dee at Glenlochar, holding back the Rivers Dee and Ken, thus enlarging Loch Ken. The barrage was fitted with six free roller sluice gates 45 feet wide and ten feet deep, which regulated the storage of water in the loch. The gates were made by the Kilmarnock-based hydraulic engineers Glenfield and Kennedy. These were boom times for such firms, Glenfield also made the gates for the Dunalastair intake dam on the Grampian scheme (page 27) and installed sluices and valves elsewhere on these schemes and others around the world. Before construction of the Glenlochar barrage, Loch Ken was about five miles long and a half to three quarters of a mile wide, in effect a widening of the River Ken above its confluence with the Dee. With the barrage in place, the loch was extended to nine miles in length, but despite environmental concerns at the time has since become recognised as a wetland of international significance.

Downstream from the barrage, one more concrete arch gravity dam was built to impound the River Dee in a head pond and from there a 1300 yard long aqueduct carried the water to the Tongland Power Station just outside Kirkcudbright. It is seen here in picture taken shortly after being commissioned in 1935. A classically simple art deco building, its appearance is made more distinctive by the large cylindrical surge tower alongside: surge towers (also referred to as chambers or tanks) acted like safety valves when the water flow was too great, and reservoirs when too slight. Equipped with three vertical turbines of 11MW capacity each, operating under a head of 114 feet, the station made the largest contribution to the scheme's initial overall capacity of 102MW. The designer of the Galloway scheme, William MacLellan, who hailed from Palnackie, died in 1934, just prior to completion of the work, and is commemorated by a plaque at Tongland. In modern times, the Galloway Hydros Visitor Centre at the station has become a tourist attraction.

Dams, tunnels, channels and pipes were the essential components of schemes designed to generate hydro-electricity from a high head, as distinct from a run-of-river system. The tunnel shown here was part of the Grampian Hydro-Electric Scheme, the first large-scale project intended to harness Highland water resources for the public supply. A dam raised the level of Loch Ericht by fourteen feet and from there the water was drawn off into the horseshoe-shaped tunnel. Nearly 5,000 yards long, twelve feet, four inches in diameter and lined with aluminous cement, or 'Ciment Fondu', it fed water to the valve house, and then the pipeline, which dropped almost 500 feet to Rannoch Power Station (front cover). There the three largest vertical generating sets made in Britain at that time went into operation in 1930. In 1953, the Grampian scheme was extended to the west of Rannoch by the North of Scotland Hydro-Electric Board when they commissioned the 6MW Gaur Power Station, the first fully automatic hydro-electric station in Europe.

Having passed through the Rannoch power station, the water was discharged into Loch Rannoch, a body of water occupying an area of over seven square miles and, with additional water being fed in from Loch Ericht, one that represented a substantial reservoir for hydro-electric generation. At its eastern end, adjacent to the village of Kinloch Rannoch, a weir was constructed to control the volume of water flowing out of the loch into the River Tummel. Five miles down river a new loch, or reservoir was formed behind the impressive looking intake dam at Dunalastair, built of concrete on a rock foundation and seen here during construction. The distinctive central 'goalposts' contained two 25 feet square sluice gates designed to allow the passage of floodwater, a not infrequent occurrence in these parts. Just above the dam on the south bank, water was taken off through screens into a three-mile long concrete-lined aqueduct leading to the second of the Grampian scheme's generating stations at Tummel Bridge.

Cut along the hillside and varying between fifty and eighty feet wide, and eleven and fifteen feet deep, the aqueduct from the Dunalastair dam conveyed the water on a shallow gradient to a terminal known as a 'forebay'. Here the water passed through screens into two steel pipelines of twelve feet six inch diameter and dropped 170 feet to the Tummel Bridge power station. It went into operation in 1933 with two horizontal generator sets. After driving the turbines the water was discharged back into the River Tummel, which continuing its descent from Dunalastair is seen in the foreground of this picture with the power station and feeder pipes behind. The Grampian hydro scheme was in the early stages of planning and construction when the Central Scotland Electricity Scheme was drawn up in 1927 and although outside the area, the output of the two stations, when they came on stream, was fed into the newly established grid through a sub-station at Abernethy.

Tom Johnston, a staunch Labour man from Kirkintilloch, served as Secretary of State for Scotland in Winston Churchill's wartime government. He and the Prime Minister were political opposites who had mutual respect for each other and so, when Johnston proposed a huge programme of developing hydro-electricity in the Highlands he had Churchill's support. Parliament backed him too and in August 1943 passed the legislation that brought the North of Scotland Hydro-Electric Board into being. A year later, plans were published for the first three projects, Morar, Lochalsh and the big Loch Sloy development. The two small schemes were commissioned on the same day in December 1948, but although work on Loch Sloy started in June 1945 it was delayed by dreadful weather and the post war shortage of materials. The scheme involved the construction of tunnels, aqueducts and a huge buttress dam to greatly increase the size of the small loch and thus create a 900 feet head above a new power station situated on the shores of Loch Lomond at Inveruglas, and shown in the picture on page 2. Commissioned in 1950, it is featured here in a contemporary advertisement for one of the principal contractors.

With Loch Sloy under way the Hydro Board promoted the Tummel-Garry Scheme; a major expansion of the Grampian Electricity Supply Company's 1930s developments at Rannoch and Tummel Bridge. The upper element of the new works consisted of a dam on the River Erochty to form a reservoir that was also fed by a twelve-mile-long tunnel from the Bruar Water and upper reaches of the River Garry. Compensation water from the dam drove the small Trinafour station, but the principal purpose of the reservoir was to divert water through a six-mile-long tunnel to the 75MW Errochty Power Station. After use, the water was discharged into Loch Tummel, which was enlarged and deepened as a reservoir by another mass concrete dam built two miles downstream from the loch's original outflow into the River Tummel. From there a further tunnel and pipeline fed water to the 61.2MW Clunie power station, seen here nestling in the Tummel Valley. The site was at the former confluence of the Garry and Tummel Rivers, which became part of the artificial Loch Faskally formed by the construction of the last of the scheme's dams at Pitlochry.

The flooding of these wild landscapes was not universally welcomed. Pitlochry, with its tourist hotels and shops selling an image of the Highlands didn't relish an invasion of navvies, while landowners feared the loss of fishing and shooting, and the coal industry wanted to maintain its monopoly of fuelling power stations. Despite having to contend with vociferous protests the Board stuck to their belief that in time the hydro scheme could be a tourist attraction as well as a major benefit. The Countess of Airlie, wife of the board's chairman, the Earl of Airlie, made the first concrete foundation 'stone' in April 1947. Initially, as mud, rocks and rubble scarred the countryside, the objectors might have seemed right but once completed, the mature scheme, with a visitor centre at Pitlochry, has indeed proved attractive to tourists. The Pitlochry dam and 15MW power station are seen here not long after 1951 when it and Clunie elements of the scheme were commissioned. In the foreground is the fish ladder, designed to allow salmon to migrate upstream, and be observed by visitors through a window in the ladder wall. This concern for fish and other wildlife was a feature of hydro-electric planning.

The 69MW Fasnakyle power station on the River Glass was the largest in the Mullardoch-Fasnakyle-Affric scheme, which was given the go-ahead in 1947. The Board planned to harness the resources of the River Beauly basin, to the west of Inverness in two separate developments. The one feeding Fasnakyle involved dams at Loch Mullardoch and Loch Benevean, a tunnel with a small power station between the lochs, and a tunnel from Loch Benevean to feed Fasnakyle. Care was taken in the design of the system to minimise the impact on Loch Affric and the River Affric and the power station was faced with sandstone from Hopeman, near Burghead to soften its appearance in the landscape. To the north in Strathfarrar, two dams were built to contain Loch Monar, one a mass concrete gravity structure, the other a double curvature arch that required less concrete. A further dam contained Loch Beannachran and the system was linked with aqueducts, tunnels and two power stations. Two more small dams and power stations on the River Beauly completed the scheme.

Further north, to the west of Dingwall, work on the scheme to harness the River Conon and its tributaries began in 1946 and continued in three phases for fifteen years. In the first phase Loch Fannich was dammed and a tunnel cut to divert water to Grudie Bridge power station. Phase two involved construction of dams at Vaich, Glascarnoch, Fannich, Meig, Luichart, Torr Achilty, Loch Droma and Loch a'Chuilinn, and four generating stations Achanault, Mossford, Torr Achilty and, the one shown here, Luichart. Built using local stone from Tarradale and situated at the junction of the Rivers Meig and Conon, it was the largest station in the scheme with an output of 34MW. It wasn't the first in the area; in the 1920s the Ross-shire Electricity Supply Company had built a station on the Falls of Conon between where the Luichart dam and power stations were later sited. The lowest phase of the Conon scheme involved dams to form Loch Orrin, which fed water by tunnel and pipeline to Orrin power station and in a final flourish, all the water in the scheme passed through Torr Achilty station.

The most northerly of the hydro-electric schemes was at Loch Shin in Sutherland. It consisted of three principal elements and three main power stations, the furthest north being a system of aqueducts and a tunnel that diverted the headwaters of the River Cassley to the Cassley Power Station on the shores of Loch Shin. The level of the loch was raised about 108 feet by a concrete and embankment dam built at Lairg where the loch narrowed. The Lairg power station was incorporated into the dam structure in a similar arrangement to that at Pitlochry. The water that passed through the power station was gathered in a lower reservoir, Little Loch Shin, which was contained by the dam shown in this picture. The dam diverted a flow of water into the river, controlled the level of the loch and fed a five-mile long tunnel that also took a feed from the River Grudie. The tunnel, in turn, fed the Shin power station at Inveran.

Completed between 1956 and 1961, the Breadalbane scheme was made up of three separate elements. The first, on the north side of Loch Tay, gathered water from numerous streams into Lochan-na-Lairige, which was formed behind a 137 feet high mass concrete buttressed dam. It was 1,400 feet above Finlarig power station, the highest head for any of the Hydro Board's stations. To the west, another system of aqueducts, tunnels and dams fed water to the 45MW Lochay Power Station passing through two small stations on the way. On the south side of Loch Tay, another series of tunnels, aqueducts and dams filled the new Loch Lednock behind a dam designed to take account of a history of earth tremors in the area around Crieff and Comrie. Linking up with other channels, the water was fed down to a power station housed in an underground chamber at St. Fillans. The outfall from there went into Loch Earn and flowed on to this weir at the eastern end of the village, which diverted water into a channel to feed the small Dalchonzie station.

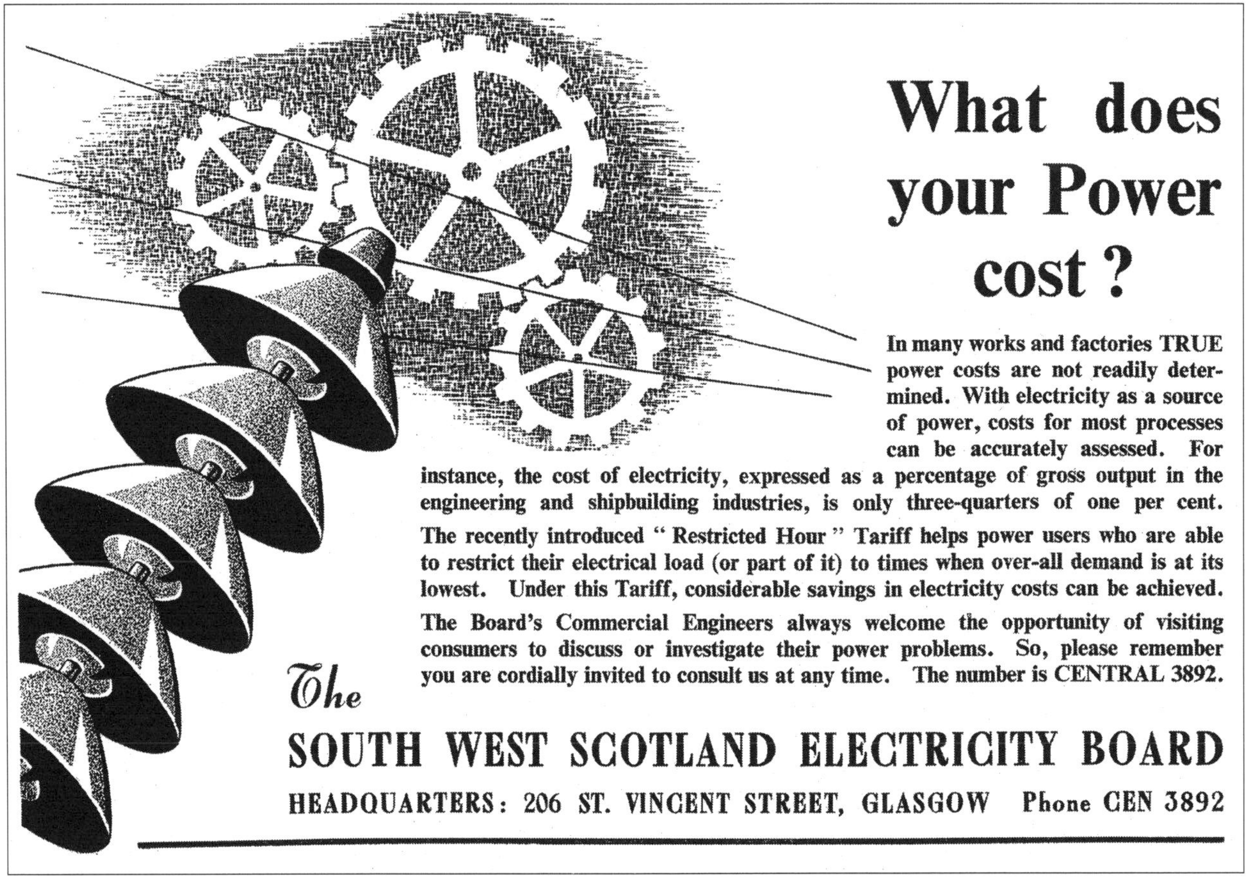

Tom Johnson did not stand for parliament in the 1945 General Election, which the Labour Party won by a landslide, but instead chose to remain chairman of the North of Scotland Hydro-Electric Board. That was crucial to securing the Board's independence when his old party colleagues in government embarked on a programme of nationalisation, including the Electricity Supply Act of 1947, which brought generation and transmission under state control: Vesting Day was 1st April 1948. Britain was divided into fourteen areas, each with a divisional controller responsible for generation through the new British Electricity Authority (BEA), and an area board for distribution. In Scotland, with the Hydro Board remaining as a separate entity in the north, two area boards were set up to cover the south east and south west. The industry faced a big challenge. Generating capacity, hard-worked during the war, was inadequate, inefficient and in many cases worn out and in need of replacement. On the supply side, outages were frequent, people were encouraged to self-ration and demand – especially in farming areas – was growing rapidly. A hint of the industry's post-war difficulties is evident from this advertisement.

To meet the growing demand, the BEA built over 40 large, new coal-fired power stations across Britain; using coal was important for a government that had also just nationalised the mining industry. One of these stations was at Braehead near Renfrew on the south bank of the Clyde (*left*). The first 100MW phase was commissioned late in 1951, quickly upgraded to 200MW and the station's capacity was increased by another 60MW in 1961. Existing stations at Clydesmill and Portobello were also upgraded by 60MW each. Compared to the hydro stations, Braehead was big but didn't endure as long and was closed in 1979 with the site later redeveloped for that era's growth industry: shopping. The next major piece of legislation was the Electricity Reorganisation (Scotland) Act of 1954, which replaced the short-lived South West and South East Electricity Boards with a single South of Scotland Electricity Board (SSEB). No longer constrained by area boundaries, they built a 760MW coal-fired generating station on the north shore of the Forth estuary at Kincardine, Fife (*below*). The site was close to known coal reserves, but had the added advantage of being where the transmission lines of the existing grid crossed the river, connecting the older stations at Dunfermline and Bonnybridge. The towers were already a feature of the area, 338 feet high, supporting a span of 3,050 feet and giving a clearance over high water of 158 feet.

Bigger power stations presented big challenges associated with getting the generating equipment on site. In January 1956, C. A. Parsons of the Heaton Works, Newcastle-on-Tyne won the contract to supply three 120MW turbo-generators for Kincardine. These were precision-engineered machines, very heavy and easily damaged by careless handling, and between the northeast of England and Kincardine there was plenty of opportunity for that. Roads were inadequate, there were no motorways and the generators were too big to go by rail, so they were shipped from the Tyne to Rosyth where there was a large enough crane to lift them. For the last few miles of the journey the machine was in the care of the road transport firm Pickfords, then part of the nationalised British Road Services, but weight and other restrictions on bridges made it impossible to take a direct route, so instead of going west, the low loader, with two tractor units in front and a third at the rear went north out of Rosyth.

Crawling along at not much more than walking pace, the convoy headed for the narrow medieval streets of Scotland's ancient capital, Dunfermline. The modern roads around the town would have made the journey less tortuous, but in the late 1950s the main route was along High Street and down to the sharp right turn into Chalmers Street at the gates of Pittencrieff Park. If that turn was tight, the left turn into Pittencrieff Street was even tighter and, as this picture makes clear accomplished with fractions of inches to spare. The Parsons generators were delivered singly, the first commissioned in December 1958, the second in September 1959 and the third in time for the official opening by the Queen in October 1960. The station was brought up to its planned capacity by 1963 with the installation of two more 200MW generators. With twin 400 feet high chimneys it became part of the local landscape, but was decommissioned in 1997 and subsequently demolished.

Kincardine's boilers could use low-grade pulverised fuel, welcome news for the new National Coal Board (NCB), which hadn't just acquired the collieries, but also vast accumulations of waste material. Some of this, in the form of pit bings, was plain for all to see, but less visible were the significant deposits of slurry in the settling ponds of coal washeries. Indeed, many people may have been surprised to discover that dirty black coal was washed at pits before being despatched. Working in collaboration with the NCB, the SSEB built a generating station adjacent to Barony Colliery in Ayrshire, specifically designed to burn low-grade coal of varying quality dredged out of slurry ponds. When it arrived at the power station the fuel was a glutinous sticky mass that had to be dried and treated before use. With an installed capacity of 60MW, Barony wasn't big, but it did clean up a lot of the coal industry's mess, and put it to good use. The station, seen here, opened in October 1957 and remained operational until 1989.

Having demonstrated at Barony that it was practical to generate electricity by burning colliery waste, the SSEB built another station to utilise the material built up by Fife's coal industry. Situated at Methil on the shores of the Forth, construction of the new station began in 1962 and the first generating unit was commissioned in 1965. As at Barony, a varied mixture of sludge-like material arrived at Methil and was fed into a large steel cylinder containing rows of flailing arms. Known as a beater mill, this reduced the coal slurry to the consistency of talcum powder, which was blown into the furnace. At full capacity Methil burned around 1,000 tonnes a day, producing, in the process about 240 tonnes of ash, most of which was collected from the flue gases before reaching the chimney while some fell into the boiler ash pit. The ash was sold to make concrete products. After consuming most of Fife's slurry, the station sourced material from other pits and so between them Methil and Barony contributed to a massive environmental clean-up.

Worn out after two world wars and two decades of labour troubles and economic woes, the coal industry was in a poor state by the time it was nationalised in 1947. To improve its position the National Coal Board set about developing major new collieries two of which, Bilston Glen and Monktonhall, were in the Lothians. With the prospect of a large tonnage becoming available in the early to mid 1960s, the South of Scotland Electricity Board planned a new 1200MW thermal power station at Cockenzie, on the East Lothian coast. Designed by architect Robert Matthew of Johnson-Marshall & Partners, construction began in January 1962 and was completed for a formal opening in May 1968. The coal, crushed into lumps of about an inch, was delivered to the power station by so-called merry-go-round trains of 900 tonnes capacity, one of which is seen here with the extensive coal storage area to the left. Without stopping, the hopper wagons were unloaded automatically into bunkers as they passed slowly through the discharge house. Two 460 feet high chimneys gave the station a distinct identity, but they were demolished along with the rest of the structure following closure in 2013.

Not content with one super-sized station, the SSEB planned another at Longannet, on the Forth shore to the east of Kincardine opposite the Grangemouth refinery. That proximity meant that oil firing was a real possibility, so a nervous coal industry came up with a radical scheme to supply the station from a series of mines working a seam known as the Upper Hirst. The plan involved connecting the mines by a tunnel fitted with a conveyor belt that would carry the coal underground for over five miles before emerging at the power station; a major feat of engineering. With an installed capacity of 2,400MW, Longannet, seen here under construction, opened in 1970 and was, at one time, the largest operational station in Britain. It was also, along with Kincardine, Methil and Dunfermline, one of four stations in Fife, which taken together with Portobello and Cockenzie across the water, made the Forth estuary Scotland's powerhouse, but these stations all burned fossil fuels and by the end of the 20th century concerns were growing about their environmental impact. There were hopes that clean coal technology, or carbon capture and storage could be developed at Longannet, but despite its size and importance, it too was shut down in 2016.

After the Second World War the requirement for military plutonium led the United Kingdom Atomic Energy Authority to initiate research into ways of adapting nuclear energy to the production of electrical power. The decision to build a prototype nuclear power station at Calder Hall on the Cumbrian coast was taken in February 1953; it was formally opened in October 1956. Before that, the Minister of Fuel and Power announced to the House of Commons that its two reactors would be increased to four and that another station, also of four reactors would be built elsewhere. The site selected was a former RAF training airfield at Annan in Dumfries-shire. Named Chapelcross, the station was officially opened in May 1959 with four gas-cooled 'Magnox' reactors capable of driving generators with an output of 240MW. The power produced by Calder Hall and Chapelcross was effectively a by-product of plutonium production, but the two plants also put Britain at the forefront of nuclear power generation and the birth of a new industry. Chapelcross was taken over by British Nuclear Fuels Limited in 1971 and decommissioning began in 2004.

The entire length of the Scottish mainland separates Chapelcross on the Solway coast and the United Kingdom Atomic Energy Authority's other research establishment at Dounreay in Caithness. The facility was set up on another former RAF station in 1955 and eventually consisted of three reactors started up successively in 1958, 1959 and 1977 to test different technologies. They closed in sequence in 1969, 1977 and 1994, and although not primarily built as power stations did contribute significant amounts of electricity to the national grid. The facility also made a big contribution to the local economy of Thurso. When Dounreay was set up, nuclear power was seen as the cleanest and most cost-effective way to generate electricity, but over time the potential hazards made it politically toxic. The paradox is that Dounreay was also the country's most easily recognisable structure associated with power generation thanks to the 135 feet diameter steel sphere. Constructed over two years by Motherwell Bridge, and three times bigger than any similar structure built in Britain before the 1950s, it was another major engineering triumph.

Demand for electricity was effectively doubling every ten years by 1955; coal supplies were uncertain and hydro insufficient, and although no other country had by this time built a civil nuclear power station the government announced plans for three, two in England and one to be built by the South of Scotland Electricity Board. The site had to be level with a firm, preferably rock foundation, close to a source of cooling water at near level to avoid pumping, and with access to substantial fresh water for boiler make-up. Hunterston, a scenically attractive spot on the Ayrshire coast was selected and after a public enquiry the go-ahead was given in July 1957. The first reactor came on stream in September 1963 and the second the following February. With a net output of 320MW it was big, but not big enough and so a second station, Hunterston B was built alongside, also with two reactors generating 1,320MW by the early 1970s. It was still not enough, and continuing with the nuclear option, the SSEB selected a site for a new station of similar size at Torness on the East Lothian coast. The plans, exhibited at Dunbar early in 1974, provoked considerable opposition and a public enquiry was held in June. Although the Board prevailed, that didn't stop the dissent and in May 1978 thousands of people marched in protest to the proposed site, but the SSEB, faced with having to meet an increasing rate of demand started to build in 1980. Based around two advanced gas cooled reactors (AGRs), the station began generating power in 1988. On privatisation Torness passed to Scottish Nuclear and in 2009 to the French state-owned company EDF. It is expected to remain operational until 2030.

Atomic Power for the World

With several twin-reactor power stations under construction Britain's nuclear power programme has reached an advanced stage. This station, which will have an installed capacity of 360 MW, is being built by the G.E.C. at Hunterston, near Glasgow, for the South of Scotland Electricity Board.
The General Electric Company Limited of England has immense resources in men and materials and is prepared to build nuclear power stations anywhere in the world. In its Atomic Energy Laboratories, scientists and engineers are constantly engaged upon research and development work on every aspect of atomic energy for peaceful purposes.

G.E.C. *Atomic Energy Division*

THE GENERAL ELECTRIC CO. LTD, MAGNET HOUSE, KINGSWAY, LONDON, W.C.2

A264

Storing electricity in large volumes is a continuing challenge, so the supply industry has to balance usage, topping up the base-load units with others that can be brought on stream quickly when needed. It is also difficult to reduce the output of big stations, especially nuclear ones at times of low demand, a conundrum that the Cruachan pumped-storage hydro-electric scheme in Argyll was designed to resolve. High above Loch Awe on the side of Ben Cruachan, and seen here on a sunny day in 1971, a mass gravity buttressed dam 1,050 feet long and 150 feet high was built to create a large reservoir. Below it, a huge cavern was excavated out of the rock to accommodate four 100MW reversible turbines. A barrage across the River Awe fixed the level of the loch so that, during the night surplus power could be used to pump water up through the turbines to fill the mountainside reservoir. The station could then generate power during the day by reversing the system and letting the water drop back down through the turbines and out to the loch. This ingenious way of storing electricity was formally opened in 1965 by the Queen. Since then, marketed as the 'Hollow Mountain', it has become a tourist attraction on the A82 Oban road.

After nationalisation in 1948 the North of Scotland Hydro-Electric Board kept the name, but also operated conventional stations like Dundee's coal-burning Carolina Port. Built in 1909 to replace smaller stations that earlier provided power to the city and its trams, its output was a modest 70MW so the Board decided to build a new station, Carolina Port 'B'. With limited coal storage space on site, this would be the first public-supply station in the country to be fired by oil and, with clean lines and a 350 feet high chimney, a dramatic addition to the riverfront. It came into service in November 1965 and in six out of the seven following years beat all other steam stations in Britain for thermal efficiency, but political clouds were gathering, which another oil-burning station came to symbolise. The South of Scotland Electricity Board, with a portfolio that included coal, hydro and nuclear, chose to diversify further with a giant oil-fired station at Inverkip on the Clyde Coast. Construction began in 1970 and when the miners went on strike in 1972 it must have seemed like a good decision, but in October 1973 oil-exporting countries in the Middle East cut supplies in a six-month-long politically motivated embargo. Prices increased four fold. At the same time, another prolonged miners' strike led to a three-day working week, a general election and a change of government. Although the dust settled, oil prices continued to rise rendering Inverkip uneconomic, its three 676MW generators used only occasionally to meet peak demand. The station came into its own during the miners' strike of 1984/85, but generation ceased in 1988 and the station was mothballed. It was decommissioned in 2006 and the 778 feet high chimney, briefly a coastal landmark, was demolished in July 2013. On the banks of the Tay, Carolina Port 'A' closed in 1977, the 'B' station followed in 1982 and the chimneys of both were demolished within a week of each other in April 1984.